Ghost Hunting in Colorado

Theories for World-Wide Investigators

Clarissa Vazquez

This book is dedicated to ghost hunters and paranormal investigators throughout the world – without your research and dedication, we would not have the knowledge we have today, nor the discoveries of tomorrow; To Cindy Hines and the Frontier Historical Society for the wealth of information; To my friends and family for putting up with yet another book project; Of course, to my Jacob for being a constant inspiration. I love you!

C.V.

Table of Contents

FORWARD
By Christopher Balzano

It is a weird thing we do, those of us who look for ghosts. You can call us ghost chasers, ghost hunters, or paranormal researchers (my personal choice), but the perception of who we are and what we spend our time doing doesn't change too much over time. People may be willing to watch television shows about it, and, according to polls published every few months, most believe ghosts are real and have had their own experiences. That doesn't mean they think differently about people who go out looking for them. See, believing in ghosts is okay, but spending your time trying to find them is still just slightly south of normal.

I've seen why time and time again. I have spent the last four years studying those who study the paranormal. While working on my book "Picture Yourself Ghost Hunting" a few years back, I sent out a questionnaire to investigators asking their opinions on looking for ghosts and some basic ideas behind what they do and how they feel. Since then I've had the chance to meet many of them at conferences and observe them doing what they do. I have called several in to get background information or asked them to hit the field in an area of the country I can't get to.

We are an odd bunch.

Many of the ghostly obsessed I come into contact with spend time not looking at the underlining conflict in what they do.

They devote time to gathering evidence, yet almost all admit the only thing that will convince someone who does not believe is to have a personal experience. They avoid the primary question I opened that book with; why are you doing this? Many can give a cursory reason (to understand something that confuses most people or to try and understand what happens to us after we die) but that is not really the heart of the question. I am not asking the spark for your hunt but rather what you expect the end result to be.

To me, looking into the unexplained is not about trying to prove anything. I have sat in a house and heard an unseen voice call my name and stood in the woods as a dark figure began to form in front of me as the flashes of light surrounded the group I was with. I have held the woman's hand as she explained the person I saw was indeed her dead father. I don't need to prove anything to people who don't believe or who refuse to look. I'm getting too old to shout at concrete walls. So if I have nothing to prove to others or myself, why continue to look?

Think of a wake. Popular psychics say most people attend their own funeral, but if we step aside from that, we have the nature of our modern, Western burial rights. Everything is based on holding the hand of the living, not the dead. As an Italian I can tell you the dead are remembered at these events in bursts, but the heartbeat of the occasion is the celebration of life and family. We comfort, but we also plan our weekend and gossip and remember and push aside conflict for a while, although not long enough to forget there is one. Death is a chance for the living to take a breath.

The search into the supernatural, walking head-on to the paranormal, has always been for me about the human light of life, either living or dead. It's about the ones left behind and the glowing ember of essence left over. It's about the family who can't deal with their dream home and the sister who gets a message from her dead brother about a missing birthday present. It's about the woman who called me in those early days and asked me why God was punishing her with a ghost who kept waking her daughter up and hiding her pearls. The science of ghost hunting, which I embraced more in the past than I do now, reinforces, but it also clouds. What is the use of measuring the level of magnetism in a place, which we can't even confirm means anything, if your only motive is to create an argument for something? Scientific evidence, as it now stitches its way into paranormal investigating, is defeated if the investigator tells the man he has proven the shadow he sees every night coming out of his closet is not really there.

A balance has to be struck. Clarissa Vazquez does that.

I first contacted her while working on a book about ghost photography and was immediately impressed by her understanding and her ability to use field work to underline a journey into something deeper. Shortly after, I asked her to appear on a radio show I work with called Spooky Southcoast. She came on with almost no notice; something I discovered was a very Clarissa thing to do, after a day of teaching classes and with no time to prep. There she stood between me, the analytical folklorist, and one of the co-hosts, a Harvard-trained scientist. What unfolded was an excellent debate (I highly recommend you find it somewhere online) where I disagreed on just about everything she said. Agreement wasn't really the point though. The intellect she showed and her ability to communicate those ideas won me over. Moreover, it became clear she had the mind of a giant in the field.

Evidence without humanity is good for creepy material on a Web site or a two minute puff piece on your local news around Halloween time. You understand nothing. Asking the next level of questions, or even thinking about what those questions might be, should be the work of paranormal researchers. Otherwise you're just a ghost hunter.

Investigation is safe in the hands of Clarissa. At its best, looking for ghosts is a combination of history, science, social science and psychology, folklore, and anthropology. You'll find those through these stories, as well as that attention to the goose bump factor that comes from telling a good ghost story. Some of the stories are from places you know, especially if you're from Colorado or are familiar with her work. That doesn't really matter. It's not as much seeing the place as seeing it through her eyes. It's not about understanding something as much as it is seeing the possibilities she lays out. It's about the tour guide as much as it is about the tour, and you're safe in her hands.

The president of the PTO where I worked once approached me and said she heard I was teaching the students about Satan, She had heard about my work and assumed looking for ghosts meant I also sacrificed small children and bit the heads off bats, and she was concerned about what I would be teacher her daughter. I assured her I was not evil, and she gave that look people give. It's the same look people give when they hear people spend thousands of dollars to go to a science fiction convention only mixed with a touch of fear and a bit of pity. I get it a lot. A few days later she came into my classroom and asked me to explain what she thought might be a ghost in a picture she had taken.

That's the nature of what we do. People view it with a morbid obsession they can't explain and hate to admit. Of course, they always have a story afterwards. So we'll keep on truckin, doing what we do for whatever reason we do it, and they'll continue to listen to the stories we have to tell. It's an old folk belief we have to look for ghosts in the dark, but it's good to know Clarissa Vazquez is doing what she does in the light of day.

Christopher Balzano

Content Director, Spooky Southcoast: www.spookysouthcoast.com and author of *Picture Yourself Ghost Hunting, Picture Yourself Capturing Ghosts on Film,* and *Ghosts of the Bridgewater Triangle*

What Makes Colorado Unique?

A Bit of History

The name *Colorado* comes from the term "colored red" given by the Spanish explorer Francisco Vasquez de Coronado when he explored the area in 1541. The area was originally split between France and Mexico, with the United States gaining part in the Louisiana Purchase in 1803 and the rest as a result of the Treaty of Guadalupe Hidalgo in 1848.

Even before the European procurement of the area, Colorado was home to the indigenous peoples of the Apache, Arapahoe, Crow, Comanche, Cheyenne, Hopi, Navajo, Sioux, Ute, and other Native American tribes. They were able to live comfortably in the mountains as well as the plains where they could hunt and gather food.

Sadly, as was the case with the indigenous peoples of the eastern seaboard and the Great Plains, the native tribes of Colorado were relocated to reservations in Oklahoma and Utah. Some went peacefully, while others resisted. The "Colorado War" between the United States and Arapaho, Cheyenne, Comanche, and Kiowa tribes of the planes lasted from 1863 to 1865. It included the famous Sand Creek Massacre in 1864 and other battles that resulted in the relocation of the tribes to present day Oklahoma.

In 1858 - just prior to the native relocation, gold was discovered by William Green and John Beck in the area along the South Platte River, now known as Denver. Thousands of people flocked to the area seeking their dreams and fortune. Some were successful either in the mines of gold, silver, zinc and lead, or as entrepreneurs - opening businesses that catered to the mining communities that were sprouting up all over the territory. Most people however, were not successful and those that remained settled as farmers or ranchers.

February of 1874 brought with it the beginning of the Alfred Packer expedition. Six men journeyed deep into the San Juan Mountains near the present day town of Lake City for the purpose of prospecting. In April, only Packer returned, claiming the other five had died in various ways – at least one by his own hand. He had resorted to cannibalism, eating his companions in order to stay alive. Alfred Packer was ultimately arrested and convicted of murder. He was initially sentenced to death, but the charges were later reduced to manslaughter with a subsequent sentence of forty years in prison. Stories of the Colorado Cannibal are told all over the country!

Colorado became the 38th state of the union in August of 1876. It was dubbed "The Centennial State" by President Ulysses S. Grant since statehood was granted just 28 days after the 100th birthday of the United States.

On September 29, 1879, Colorado witnessed the Meeker Massacre and subsequent Battle of Milk Creek, which are considered to be the last major Indian uprising. Nathan Meeker and seven of his associates were killed by members of the Ute tribe because of a dispute over a race track. Major Thomas Thornburgh and nine of his men were killed at Milk Creek en route to provide assistance to Meeker. The Utes were eventually relocated to a reservation in Utah.

The twentieth century saw its fair share of historic moments as well. Gunfighters and lawmen came and went, the Ku Klux Klan began making appearances in the 1920's, race riots sprouted throughout the metropolitan areas in the late 1960's, and the tragedy at Columbine High School in 1999 – each having left their marks on Colorado's history.

With such an eventful past, one could be left to wonder what could possibly be in Colorado's future!

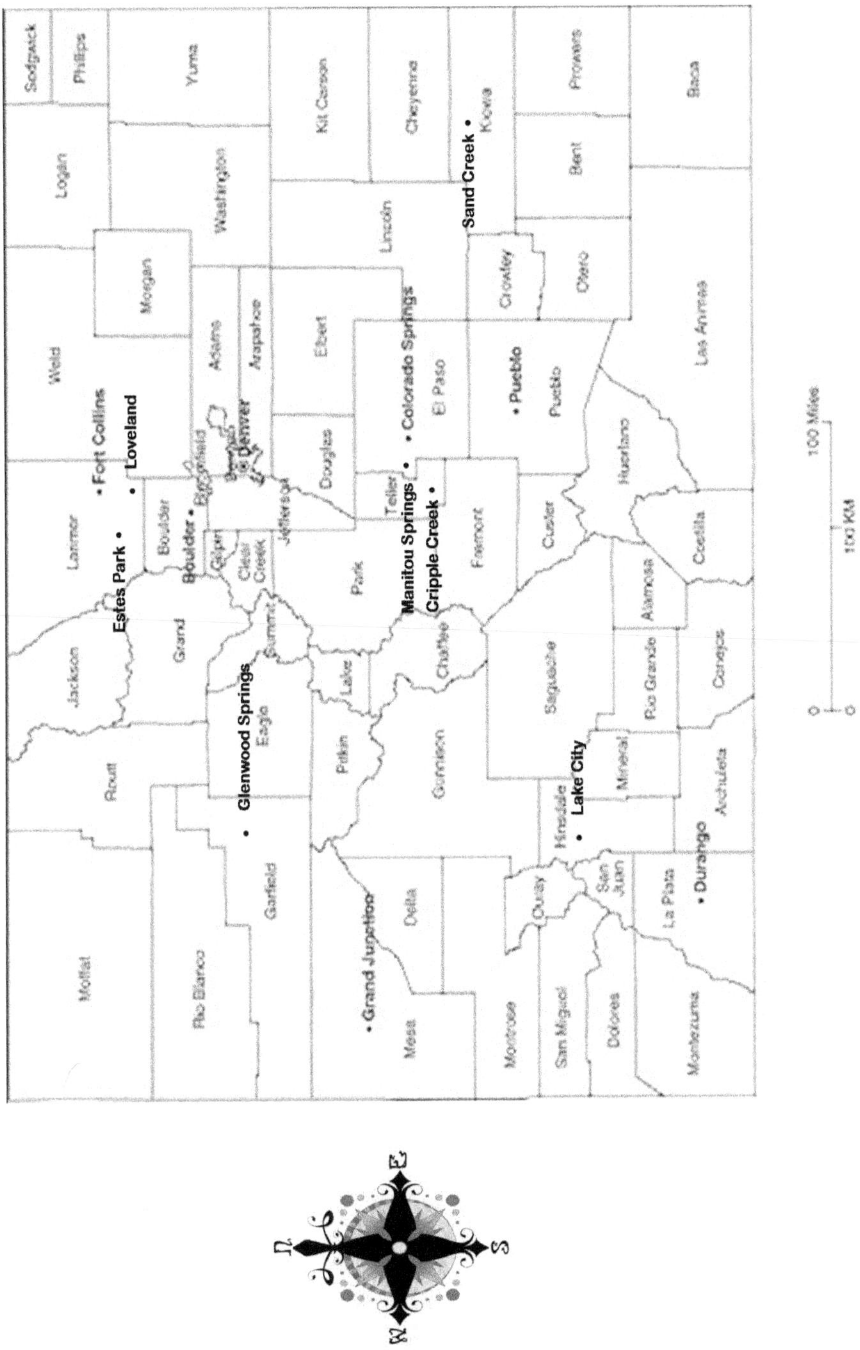

Fort Collins
Loveland
Estes Park
Boulder
Denver
Glenwood Springs
Grand Junction
Manitou Springs
Cripple Creek
Colorado Springs
Pueblo
Sand Creek
Lake City
Durango
Sedgwick
Phillips
Logan
Yuma
Washington
Morgan
Weld
Larimer
Jackson
Routt
Moffat
Rio Blanco
Garfield
Eagle
Grand
Boulder
Adams
Arapahoe
Kit Carson
Cheyenne
Lincoln
Elbert
Douglas
Jefferson
Clear Creek
Summit
Park
Teller
El Paso
Kiowa
Crowley
Prowers
Bent
Otero
Baca
Las Animas
Pueblo
Fremont
Custer
Huerfano
Costilla
Alamosa
Conejos
Rio Grande
Saguache
Chaffee
Lake
Pitkin
Gunnison
Mineral
Hinsdale
Archuleta
La Plata
San Juan
Ouray
Delta
Mesa
Montrose
San Miguel
Dolores
Montezuma
100 Miles
100 KM
N
E
S
W

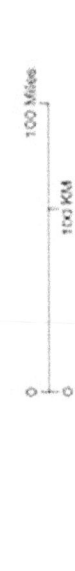

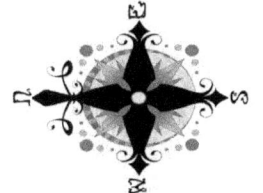

Rocky Mountain Earthquakes

When most people think of earthquakes, their minds fill with images of the Ring of Fire in Southeast Asia or of California detaching from North America and sliding into the Pacific. Earthquakes are spontaneous and random, sometimes leaving nothing behind but crumbled buildings and shattered lives for the survivors.

At first glance, Colorado does not seem like an ideal location for earthquakes. Closer inspection though shows several areas of seismic activity throughout the state. The majority of these are located in the western-central region, although the Denver area has seen its fair share of low-magnitude tremors as well.

The Colorado quakes are not attributed to tectonic plates like the major shakers in California. Instead, they are caused by the ever-changing Rocky Mountains, underground springs and rivers, and the deep drilling and mining of natural resources. This is most prevalent in the Rio Grande Valley from Leadville south to the San Luis Valley and the Great Sand Dunes National Park. Any time the earth's lithosphere is disturbed, it leaves room for settling. A coal mine, for instance, can tunnel many miles into the earth. Coal and rocks are removed and often replaced with structural supports for the safety of the mine and its workers. Without the geological "filler", the natural pressure and weight of the area have been compromised. Occasionally, a mine shaft will cave in as a result. The shock wave can be felt for several miles, depending on the location and severity of the event.

The same can be said for the areas near a natural spring, particularly hot springs where the water is so close to the earth's core, it is heated to 100 degrees Fahrenheit or higher. The higher the temperature, the higher the pressure until it breaks to the surface. With this spontaneous upheaval comes a series of shock waves that can be felt a great distance away if the pressure is high enough. This is most evident with the geysers at Yellowstone National Park in Wyoming, although Colorado has several hot springs in areas like Glenwood Springs, Idaho Springs, Steamboat Springs and Ouray just to name a few.

So, what does seismic activity have to do with ghosts?

Some researchers believe that paranormal activity intensifies directly following a seismic event. Reports of strange and unusual activity have been known to increase immediately after a geological shift. Less-than tactful investigators flock to major earthquake sites in hopes of capturing evidence of the recently and traumatically departed. This practice is frowned upon by the majority of researchers, if not condemned.

Other, more serious researchers have suggestions to contribute to the theories. One suggestion is the possibility that the events being observed are a result of after-shocks. Humans, like any other animal, are sometimes able to subconsciously detect natural events. Following a geological event like a small earthquake, our senses are going to be more in tune to our surroundings, thus providing an explanation for the rattling windows, creaking floorboards and other things not necessarily noticed on ordinary days.

Another theory is that the seismic activity disrupts the natural electromagnetic field (EMF) of the earth, thereby increasing the opportunity for paranormal activity *if* EMF is a genuine factor for ghostly encounters. Since we still do not know if ghosts produce, utilize, or disrupt EMF, further research is necessary to determine which of the theories is appropriate.

Zachariah Martin provides two additional explanations for the correlation between ghostly encounters and disruptions in the lithosphere.

> "The Tectonic Strain Theory (TST) states that friction and stress on the Earth's crust, in less magnitude than what is required to generate an earthquake, may result in high level and low level electromagnetic disturbances in the form of Piezoelectricity in the sub-surface rocks in the Earth's crust.
>
> Piezoelectricity is a naturally occurring phenomenon that occurs in certain crystals, most notably Quartz crystals. Quartz actually produces a naturally occurring electrical

charge across opposite faces of the crystal, much like a battery. This is achieved through the application of physical pressure on the crystal.

If a large piece of rock were to fracture, it may produce an electromagnetic field due to a process known as Seismoelectric conversion. This is proven through history with reports of earthquakes being accompanied by measurable, although weak, magnetic disturbances.

Due to Seismoelectric conversion, it may be a good suggestion to paranormal investigation teams and paranormal researchers to examine the local faults surrounding the site of investigation. When doing so, special attention should be paid to types of rocks in the soil of the area surrounding the investigation. The rocks that should be watched for include but are not limited to granite, limestone, and basalt. Keep in mind this theory is in the development stage but is showing great potential.

Research suggests that electric and magnetic fields produced through rock strain are most commonly found near underground geological fault-lines. These cracks are signs of high levels of strain and movement in the earth's crust."

If you are interested in studying the correlation between seismic events and paranormal activity, you can find the most up-to-date seismic maps by visiting the United States Geological Survey website: http://earthquake.usgs.gov/earthquakes/states/

*Note: The U.S.G.S. does not participate in paranormal studies. The above-mentioned website is for reference purposes only.

Limestone

You're conducting a residential investigation. Things are moving along smoothly with nothing exciting to report. Suddenly and without warning, a woman in antebellum attire enters the room and sits in the rocking chair opposite you. She does not look at you, nor does she acknowledge your presence. You try to speak to her, but it is as if she can't hear you at all. Within the blink of an eye, she has vanished – leaving behind only the very faint movement of the rocking chair or perhaps a waft of perfume. The following night, you return with every piece of equipment in your arsenal trained on that rocking chair. Just as she had the night before, the woman appears and sits in the chair. She is wearing the exact same dress as the previous night and once again, she doesn't acknowledge your existence. Then she vanishes, leaving the chair empty and your heart racing! With any luck, you were able to capture some essence of her with your cameras.

Most investigators can tell stories similar to this where someone appears before them, does not interact with them, then vanishes into thin air, walks through a wall, jumps off a cliff, or some other disappearing act. The common definition for this type of phenomena is a Residual Haunting.

So, what is a residual haunting and what causes it?

By literal definition, residual is the residue of something from the past. In this case, the spirit of someone or something that was once living. Residual hauntings almost never interact with the living and frequently replay the same moment in time over and over – just as the lady appeared and sat in the rocking chair night after night. There is very little that can be done in residual cases other than study them because the phenomenon does not appear to be conscious or aware of their surroundings.

One ever-growing popular theory surrounding residual hauntings has to do with Limestone and other mineral deposits. Some paranormal investigators and now geologists believe that Limestone, Quartz and Magnetite can hold energy to be later released. Sort of like a Flintstone-era audio recorder without the prehistoric woodpecker. If the mineral deposit is large enough and the energy is strong enough, the rocks will hold on to the energy, slowly releasing it over time in short bursts. Like a CD player stuck on repeat.

There are a couple of problems with the Limestone theory - the first one being the physical components of the rock itself. Limestone is a sedimentary rock made up of a few minerals, but primarily the skeletal remains of sea creatures and coral. It is somewhat water-soluble and erodes very easily. So what holds on to the energy in the rock? Further research on Limestone deposits and residual hauntings needs to be conducted before we can know for sure if there is a direct correlation.

Quartz on the other hand can be utilized in radio and radar transmissions. Under the right conditions, it can produce an electrical charge (piezoelectricity) and transmits ultraviolet light waves better than glass. It has the physical capacity for retaining light energy and small amounts of electricity, so it would stand to reason that it might be able to hold the energy that could possibly result in a residual haunting.

The same can be said for Magnetite.

Magnetite can be found in sedimentary, igneous and metamorphic rocks. Its name comes from its ability to become and remain highly magnetized. A popular theory with paranormal researchers is that ghosts can either disrupt magnetic fields or are better-enabled to manifest in the presence of strong magnetic fields. If either of these theories is accurate, then the possibility of paranormal activity being greater in areas with large quantities of concentrated Magnetite increases significantly.

Unfortunately, there is no true test for any of these theories right now. At best, investigators can study the correlations between high areas of paranormal activity and the geology in those areas. If researchers are able to prove any one of the theories involving the geology of an area compared to the levels of paranormal activity, then we could potentially pinpoint, confirm, and identify other theories where activity is concentrated, such as vortexes and portals.

If you experience a residual haunting, you can reference the USGS Field Records Collection to determine the geological makeup of the location. http://www.cr.usgs.gov/fieldrecords/

Your haunting could possibly be attributed to the very land on which it's occurring!

Geomagnetic Activity

One thing paranormal researchers agree upon almost universally is that ghosts are merely some form of energy. There is a two-part explanation for this. The first part being that the human body is comprised of two things – water and energy. When our physical bodies die, the water evaporates, but where does the energy go? Energy cannot be created nor destroyed, so it has to go somewhere! The second part is the fact that energy fluctuations, whether electric, thermal, or magnetic, have been directly attributed to ghostly encounters.

Deep inside our planet is a core of molten rock. Scientists believe that the movement of this core creates a magnetic field. It is this magnetic field that enabled humans to travel with the invention of the compass. Paranormal researchers believe that ghosts can either tap into the magnetic field, enabling them to manifest into something the living can detect, or disrupt the magnetic field, giving the living the ability to detect them. Some "old school" researchers can tell stories about walking through cemeteries or abandoned houses with nothing but a flashlight, a camera, and a compass. During the "dark ages" of investigating, bringing a compass with you was the key because it would behave erratically in the presence of a ghostly encounter.

But what causes this erratic behavior?

Researchers have yet to determine which of the theories is the most accurate surrounding ghosts and electromagnetic fields. Some believe the ghosts emit an electromagnetic field (EMF), while others believe the ghosts disturb the natural EMF. A third theory is that the ghost utilizes the EMF as a type of fuel for manifestation. Regardless of which theory is the right one, we know for sure that there seems to be a correlation between EMF and ghosts.

This correlation has caused a rather severe problem, unbeknownst to investigators. With the introduction of EMF detectors and meters, investigators were able to take the proverbial ball and run with it, thinking they were in possession of a "ghost detector". Investigators could be found mesmerized by every blip, beep, and blink of these meters thinking they were having ghostly encounters not visible to the naked eye.

Further research and experimentation has found that while EMF meters are fun to play with, they are not ghost detectors. The best use for them on investigations is to locate the naturally occurring EMF at a location to see if witnesses are potentially suffering from hallucinations due to prolonged exposure. Occasionally an investigator will get lucky and have an EMF reading in conjunction with a temperature fluctuation or possibly an EVP.

While these instances are infrequent, they are the ones that need to be closely examined.

The natural geomagnetic / electromagnetic field is not constant and can fluctuate dramatically at times. Some researchers and geophysicists are trying to prove theories that the geomagnetic activity spikes twice per year with the most severe spike being toward the end of October. These spikes can be monitored and noted during investigations to further determine if there is a correlation between what is going on deep inside the earth and possible increases in ghostly activity.

The Kennziffer Index or K-index was introduced by Julius Bartels in 1938 as a method for quantifying disturbances in the earth's magnetic field. The Kennziffer-planetary (Kp) index is the system utilized by the National Oceanic and Atmospheric Administration (NOAA) for measuring geomagnetic fields from several locations around the world. These measurements are compiled over a three-hour period and assessed based on the K-index scale.

A geomagnetic storm has occurred or is occurring when measurements from around the world are greater than K4. In order for the correlation between geomagnetic events and ghost activity to be proven, the majority of paranormal events would have to happen within three hours of a Kp 4 event or greater.

QUIET

Kp <4
The Geomagnetic field
is not active.

UNSETTLED

Kp=4
The Geomagnetic field
is active

STORM!

Kp>4
A Geomagnetic Storm
has or is occurring

Investigators can reference the NOAA Space and Environmental Center website: N3KL.org to see what the geomagnetic field is doing during investigations. With research, we can attempt to determine if the geomagnetic field plays any role in the occurrences of paranormal phenomena.

Solar Flares

From September 1 to September 2, 1859 telegraph lines around the world suddenly shorted out and the Northern Lights usually visible in the northern-most parts of North America and the Arctic were visible as far south as Hawaii, Cuba and the Vatican. In the gold mines of the Colorado Rocky Mountains, the sky was so bright miners woke up and began preparing breakfast thinking it was morning.

As the telegraph fires were put out and the atmospheric phenomena subsided, people knew that something pretty spectacular had just happened, but what was it? Religious fanatics and soothsayers preyed on the scared and the curious, telling tales of the end of days or selling potions for protection. It was determined that a massive solar flare had created a magnetically charged cloud of plasma, called a coronal mass ejection (CME). Most CME's take about three days to reach the earth. The cloud that impacted the Earth's atmosphere in 1859 was so intense it took only seventeen hours to reach us.

The Earth has had several CME's of notable size impact the earth throughout the last century. The years 1920, 1961 and 1994 each saw solar flares of notable size, but the super storm of 1859 is the largest and most destructive on record. Even more astounding is that reports of ghostly encounters seemed to increase in the late 1800's and 1920's – almost in conjunction with the CME's.

Given that the world was heading into a rather tumultuous era in 1961 with the space race and Cold War taking center stage, people did not discuss what went on behind closed doors openly and talk of ghosts and spirits was certainly taboo. This could explain for the lack of paranormal reports during that era.

The late 1990's and early 2000's saw a dramatic influx in reports of paranormal activity. Televisions have since become inundated with programs about ghosts and the paranormal. Is this most recent influx due to increased CME-related activity or is it because social stigmas are being removed and it is becoming more socially acceptable to discuss having had a paranormal encounter?

Scientists have predicted a solar flare with the capacity to produce a CME equivalent to the 1859 flare occurring in early 2012. A CME that size has the potential to destroy satellites and disrupt the majority of communication methods for an extended period of time, causing billions of dollars in damage. Because CME's are *magnetically* charged plasma clouds, a cloud of significant size could possibly bring with it an unparalleled increase in paranormal activity if ghosts do in fact utilize electromagnetic energy in order to manifest.

The NOAA has two satellites that monitor solar activity. The Geostationary Operational Environmental Satellite (GOES) 8 and GOES 10 satellites monitor in five-minute increments. These readings are analyzed and compared over a 24 hour period to determine what the sun is doing with regard to x-ray fluctuations and energy production. The reports are then posted on the NOAA website and subsequent links. The readings you find on the worldwide web are at least 24 hours old, but have lasted at least five minutes and created a significant amount of energy.

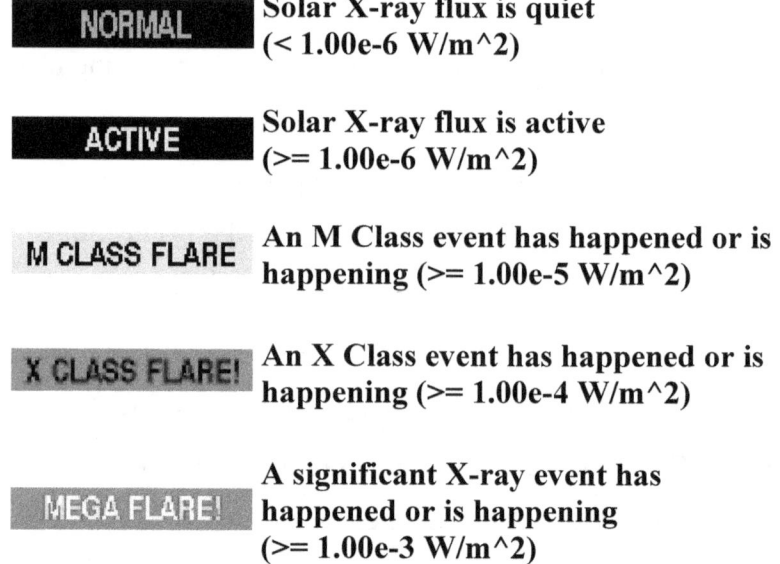

NORMAL — **Solar X-ray flux is quiet (< 1.00e-6 W/m^2)**

ACTIVE — **Solar X-ray flux is active (>= 1.00e-6 W/m^2)**

M CLASS FLARE — **An M Class event has happened or is happening (>= 1.00e-5 W/m^2)**

X CLASS FLARE! — **An X Class event has happened or is happening (>= 1.00e-4 W/m^2)**

MEGA FLARE! — **A significant X-ray event has happened or is happening (>= 1.00e-3 W/m^2)**

If ghosts utilize the energy around them in order to manifest, imagine what they could do with an unlimited resource like the sun! The solar activity can be checked on the NOAA website N3KL.org or on the homepage of many paranormal investigation organizations.

By determining the solar activity prior to or during a paranormal event, we can attempt to find a correlation between CME's and ghostly encounters.

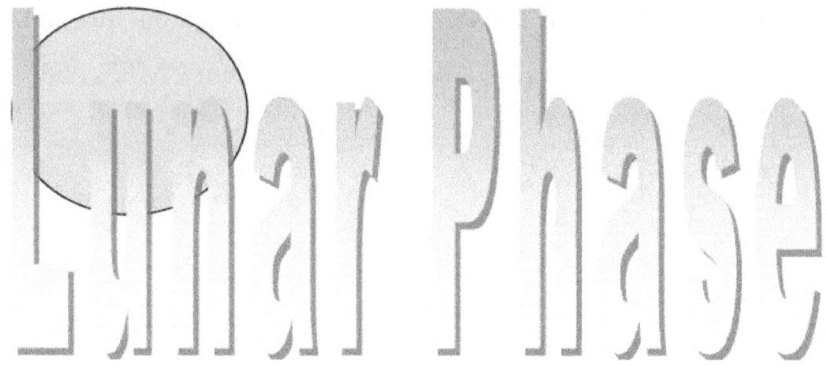

Lunar Phase

There are vast and exotic tales of frightening things happening during various lunar phases. Werewolves are said to miraculously and spontaneously grow fur-like hair and morph into some grotesque canine in the light of the full moon. Witches fly out on their broomsticks seeking innocent souls to seduce or children to eat. Cats do a tightrope act on split-rail fences while serenading neighborhoods and vampires lurk in the shadows stalking their next meal.

Stories with lunar significance are ancient, reaching back to the dawn of man. They cast shadows into our bedrooms at night and into our dreams, bringing fear and amazement to those that choose to pay attention, and even provide a source for worship in some. The moon controls the ocean tides with its gravitational pull and even determines when crops are planted and harvested. From movies to documentaries, Hollywood has utilized the moon on more than one dramatic occasion and many a young girl has fanaticized about the "man in the moon" or dreamed about her Prince Charming by its light.

In the 1960's mankind proved we were not limited to the Earth when NASA's Apollo program succeeded in putting a man on the moon. This unprecedented event changed the way we look at the moon not only from a mystical standpoint, but scientifically as well.

Scientists have long speculated that what is now the moon was once part of the Earth. A large rock roughly the size of Mars is thought to have impacted our planet during its infancy. Rounded by the Earth's gravity, the largest piece of debris became our natural satellite. It has a core and a mantle just as the Earth does along with a small amount of gravity. The moon orbits the earth every 27.3 days and for six of those days it passes through the Earth's magneto tail, causing it to become negatively charged.

For decades, scientists have been studying the correlation between the cycles of the moon and human behavior – otherwise known as the Lunar Effect. Law enforcement officials and hospital staff alike have reported increases in the services they provide during full and new moon cycles. The number of babies born under full or new moons increases dramatically compared to other times of the month. Psychiatrists and Psychologists have reported drastic changes in mental health behavior – both positive and negative – during these phases.

Zombies, vampires, and werewolves aside, theories surrounding ghostly activity are no exception when it comes to lunar phase. Just as the moon is said to effect behavior in the living, it is theorized that it affects behavior in the dead. Reports of ghostly phenomena increase during full and new moon cycles and researchers often schedule investigations and field trips during these phases with hopes of increasing their chances of finding credible evidence.

Is it the gravitational pull or the slight electromagnetic charge from having been pulled through the magneto tail that allow ghosts to become purportedly more active and visible to the living? If ghosts do in fact utilize electromagnetic energy in order to manifest, this theory would then prove to be accurate.

Contributing Colorado's elevation to this theory could provide further explanation as to why there are an increased number of ghostly reports in the higher elevations of Leadville, Central City, Estes Park, and Glenwood Springs. Because they are closer to the moon and its temporary magnetic charge, that could account for the increase in reports in those areas.

In order to further determine if ghosts are more active during the full and new lunar phases, researchers are going to have to work together to make the correlations between activity and lunar phase. The exact lunar phase is easy enough to determine with the help of a computer. Websites like moonconnection.com provide minute by minute lunar phase images and fullness percentages.

While several organizations have projects involving lunar phase, none has been found to be as thorough as the Phantom Hitchhiker Project developed by the Colorado Coalition of Paranormal Investigators (CCPI). This multi-faceted project is working toward making the correlations between ghostly activity and the theories surrounding time, temperature, lunar phase, and other factors presumed to affect ghostly encounters. The program is designed for investigation organizations around the globe to utilize (with permission) and the findings are displayed on the CCPI website. If you are interested in more information or adopting this project, contact CCPI via their website www.coloparanormal.com for details.

With continued research, investigators can better plan outings and provide a higher quality of assistance to their clients!

Psychic Trauma

In the chapters up to this point, we have discussed several popular theories surrounding causes for ghostly interactivity. Because paranormal investigation - or more specifically ghost hunting - is not yet an exact science, all we have are theories. For the serious investigator, half of the attraction to this field of research is trying to prove or disprove the theories, with the other half being the ability to provide assistance to people in need.

The final theory we're going to examine before we begin our tour through Colorado's haunted hot spots is Psychic Trauma. Not human psychic trauma, but Geographic Psychic Trauma (GPT). Countless reports of ghostly activity and ghost lore come from locations where someone or something has died. Why is this? Logic would suggest that people would want to remain with their loved ones or personal possessions as opposed to some crummy hospital, nursing home, or crash scene. So then, why are some spirits reported to remain in the location where they left their bodies?

Imagine riding in a car. You're cruising at a swift 60 mph - listening to your favorite traveling tunes when suddenly a deer appears out of nowhere. You are able to utilize your brakes and safely avoid hitting the deer - foregoing serious damage to your vehicle. During that moment of rapid deceleration, whether you felt it or not, your body was slammed violently forward at sixty miles an hour. This is easily explained by Sir Isaac Newton's First Law of Motion, also known as the Law of Inertia.

> **"An object at rest will remain at rest unless acted on by an unbalanced force. An object in motion continues in motion with the same speed and in the same direction unless acted upon by an unbalanced force."**

If you were smart, you were wearing a safety belt that kept your body from being propelled forward with enough force to cause serious injury, if not kill you on impact.

Now, imagine the same scenario with a twist. This story doesn't have the same happy ending and unfortunately, you were not able to avoid hitting the deer and subsequently lost control of your car - hitting a tree. The impact has you stunned and confused, but you are able to get out of your vehicle only to turn and find yourself still sitting in the front seat.

This type of story has been recounted by numerous near-death survivors. As far as we know, the human spirit does not have a safety belt to keep it from being violently propelled. So what happens to the human spirit when it is violently yanked from the body such as in our scenario? Paranormal researchers, parapsychologists, and even a handful of mainstream mental health professionals are starting to focus on the energy resulting from a violent death.

It is at this point in the book where we will *not* discuss the white light, heaven, or reincarnation!

When one or more spirits are believed to inhabit the scene of their violent death, it is known as Geographic Psychic Trauma (GPT). The spirit is believed to be imprinted (for lack of a better term) on the land or building where their life energy left their physical body. What researchers are learning is that the more tragic and violent the death, the more likely it will be for some form of spirit energy to remain directly connected to the place where they died. This could perhaps be attributed to the fact that humans do not have that spiritual safety belt to keep us in our living bodies.

Hospitals and nursing homes have reports of ghostly phenomena, but not to the same degree as less-than reputable mental institutions and prisons where the nature of the deaths was more likely to be tragic, resulting in possible GPT. The same can be said for crash scenes, war zones, and murder locations where the life energy was forcibly ripped from the living body.

It stands to reason that if the life energy was suddenly and violently torn from the living body, it could result in some confusion for the spirit as opposed to someone with a terminal illness who has had a bit of time to prepare for what is to come. This could in fact account for the numerous ghost sightings in the former mining towns of Colorado where thousands were lost due to tragedies such as cave-ins, explosions, and gunfights.

With the colorful history of the gold rush era in Colorado, there are many vast encounters with spirits from the gold, silver, and other mineral mines of the Rocky Mountains. Ghost lore tells of the spirits of miners eternally banging on rocks inside collapsed mines hoping for a rescue that would never come. These "tommyknockers" are also reported to be responsible for warning miners of impending doom or foretelling cave-ins, giving the living miners time to escape.

Naturally, investigating locations where a recent tragedy has taken place is not only unethical, but it is also disrespectful. It is only recently that investigators have been welcome in places like Pearl Harbor and eastern-European concentration camp sites. Locations where the GPT is recent have fresh emotional scars for the survivors and families of the departed. Sadly, so-called investigators flock to these locations, tarnishing the reputations of genuine researchers around the globe.

Anyone interested in investigating locations with possible GPT should look into their local historical archives. The Colorado Historical Society is a good start: www.coloradofieldtrip.org By collecting the historical records first, you can attempt to find out who you're potentially dealing with as well as the circumstances surrounding the GPT.

With more researchers exploring these locations, we can contribute to the theories in these locations and perhaps determine exactly why some spirits stay while others seem to move on.

Haunted Hot Spots

Hotel Colorado – Photo by CCPI

Hotel Colorado
Glenwood Springs

Built in 1891, the Hotel Colorado opened its doors to the public in 1893 with 191 sleeping rooms renting for as much as $3 a night for the luxury rooms. Proprietor, Walter Devereaux built the establishment to cater to guests near and far who wished to partake in the medicinal hot springs in the area.

The hotel has seen many changes in the last century and has been used for many things to include a naval hospital during World War II, and a retreat for notorious gangster Al Capone. Today, only 131 of the original rooms are rented out with the rates varying from $110 per night to $675 for the deluxe suites.

Hotel guests and staff are not the only occupants of the historic site. The Hotel Colorado is home to several spirits that inhabit the building. Guests and staff have reported high activity between the hours of 2-4 a.m. They have reported the elevator moving from floor to floor with no passengers, cigar smoke - presumably from Walter Devereaux - in the main lobby, and dishes being moved in the Devereaux Dining Room. A young girl in Victorian clothing has been seen playing with a ball in various areas of the hotel, and a female apparition has been reported looking over sleeping male guests.

It is rumored that the famous gunslinger Doc Holliday died in the hotel. This is not true. According to the Frontier Historical Society in Glenwood Springs, Holliday died in the Hotel Glenwood, located on Grand Avenue. At the time of his death, it was called the Hotel Glenwood Springs Sanatorium. Today, the location serves as retail and business spaces.

On the fifth floor of the Hotel Colorado, there is the spirit of a man in period clothing that many guests mistake for Holliday. Guests frequently leave whiskey and chewing tobacco for him in room 558.

During the time the hotel served as a naval hospital, a woman was caught in a lover's triangle and murdered by one of her jealous lovers. Her screams can still be heard in areas of the hotel and the room in which she died has had so much activity, it cannot be rented and serves as a storage room.

The basement contains the original hydraulic lift designed to bring supplies from the ground level such as ice blocks that were stored in the basement. It is speculated that while staying there, Al Capone used it as an escape route on more than one occasion when the hotel was raided in search of him. The lift is still operational today.

Another area of the basement served as the morgue while the hotel was being used as a Naval Hospital. Several hotel staff members have reported locked doors opening and lights turning themselves on and off in this area. No one can pinpoint which spirit occupies the basement area, although it is speculated that it is one of Walter Devereaux's many hangouts.

Paranormal activity is reported most frequently between the hours of 2 and 4 A.M. The fifth floor is an area where people have experienced a higher amount of activity. Several guests and staff have reported apparitions, strange smells, sounds, lights turning on and off, televisions changing channels, and knocks on doors with no one there. Although there were no

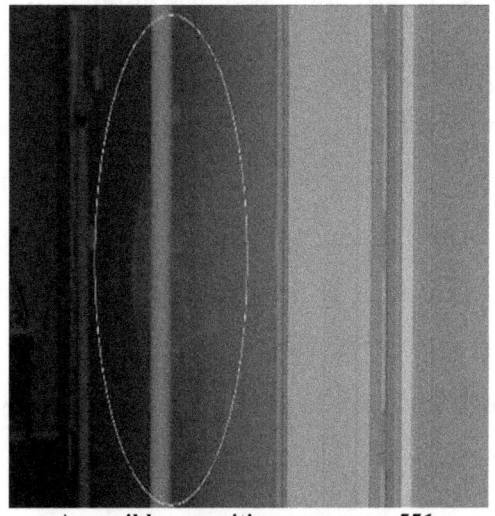

A possible apparition near room 551.
Photo by CCPI

temperature fluctuations at this location, a photograph of the area indicates what could be spirit energy near room 551 during the CCPI investigation. It was later learned that the unmarked door in front of room 551 is a door to the attic. Other guests and psychics have reported a male spirit that inhabits that area, possibly one of the workers who helped construct the building in 1893.

Room 551 itself has also had many strange phenomena. In 1982, the hotel was being renovated and new wallpaper was being applied to the walls of the rooms. The next morning, all of the wallpaper that had been applied to room 551 was rolled neatly on the floor. The wallpaper was replaced, but again found on the floor in the morning. After two or three attempts, several wallpaper samples were placed on the bed. The following morning, all but one of the samples was on the floor. The sample that was left on the bed was applied to the walls and remains there today.

The hotel has two bell towers that have been turned into luxury guest suites. Room 661 has been dedicated to one of the most famous women in Colorado – Titanic survivor, Molly Brown. Ghostly activity has been reported in this room as well as the other bell tower suite, room 662. While both suites have reports of spirit activity, the Molly Brown Suite is said to be the most active.

The phenomena frequently experienced at the Hotel Colorado reminds us that we are gaining in knowledge about ghosts and the spirit world but they still know more than we do and have the upper hand.

Miramont Castle
Manitou Springs

Jack Martinez, Contributor

Miramont Castle is an active paranormal hotspot in Manitou Springs, Colorado. It is currently a cultural museum which houses many artifacts from Manitou history. The castle was built in 1895 for an ill French priest named Jean Baptiste Francolon, as a gift from his mother. Father Francolon was a respected priest and missionary who fell ill during his travels. He was assigned to Colorado Springs with hopes of the healing waters of Manitou Springs curing him.

Manitou sits at the base of Pike's Peak, which inspired Father Francolon with its beauty. He wrote his recently widowed mother to move there. She was quite wealthy, and sent him an unknown sum of money to start construction of a home for them both.

Francolon, being an architecture fan, drew out plans for the building with reminders of the many architectural designs he experienced all over the world. When he finished the designs, the four-story, forty-six room mansion was built. Construction took over a year to complete, and included a greenhouse, servant quarters, a private apartment for his mother, a formal dining room, and a chapel. There are no official blueprints on record of the castle, which was built right into the side of the mountain.

Francolon and his mother moved into the castle in 1896, where he continued his duties for the church. A branch of the Sisters of Mercy opened a sanitarium in Manitou Springs to help the many tuberculosis patients who swarmed the city at that time, and Father Francolon assisted them in building it near the castle. During this time, it was alleged (however not proven) that he was having an affair with one of the sisters. When the alleged romance failed, the nun hung herself in the greenhouse. It wasn't very long after this incident in 1900, when Father Francolon and his mother abruptly moved out of the castle and went to New York City. After they left, there was a mysterious fire which damaged his room, and the castle was condemned. Mrs. Francolon died soon after moving that winter, and Father Francolon eventually died in 1922, all but forgetting the castle ever existed.

The Sanitarium used by the Sisters of Mercy eventually succumbed to a fire, and the Sisters decided to move to the much larger unused castle. They had it repaired and relocated the hospital in 1904. There, the nuns operated the sanitarium quite successfully until better treatments for Tuberculosis shut it down in 1928. It is well documented that hundreds of Tuberculosis patients died while in Miramont Castle.

After the sanitarium closed, the Sisters of Mercy turned the castle into a retreat for their order. This was a quite uneventful time until 1946, when it was closed and the building was converted into an apartment complex. From 1946 until 1975 the history of the castle is unclear, however it is known the castle suffered at least two fires and several people died there. A little girl died of an illness, a boy drowned in the basement, a transient was found dead inside a room, and a man was electrocuted in an accident. The building was once again condemned after the second fire and was slated to be torn down in 1975, when the Manitou historical Society decided to restore it.

The Castle was restored as closely to the original designs as people could remember, and much of the original furnishings were restored. The city furnished the rest of the castle with antique Victorian decorations from other homes and historical sites, including the original grand piano brought in by Mrs. Francolon from France.

After the restoration and conversion into a museum, many staff and visitors reported strange occurrences throughout the building. There were corroborating reports of ghosts in areas where people were known to have died, and strange reports of spirits who supposedly were attracted to the castle. It has been suggested that the spirits of Father Francolon and his Mother returned to Miramont after their deaths.

The spirit of the small boy who died is rumored to inhabit the janitor's closet in the basement. He is said to play with the toys in the Miniature museum in the basement of the castle adjacent to the closet. Staff and visitors have claimed to find items behind display cases, which are inaccessible to almost everyone, were moved overnight, and they hear the voice of a small child in the room when it is empty. He is also said to turn on the water in the sink and open the doors. There are EVP recordings of the small boy singing.

Just up the stairs from the Miniature Museum is the main lobby, where on the master staircase, the ghost of a woman in a Victorian dress resides. There have been sightings and even some pictures of her moving along the landing and going down the stairs. No one really knows who she is, or why she is there. However the fact that she is in Victorian dress indicates she was from the time of the castles construction and inhabitance.

Just to the left of the lobby is the main living room, which has a fireplace that was cut right into the side of the mountain. Many strange reports come from that room. Stories of a man dressed like a transient who wanders around the room, and the smell of cigar smoke. Staff report personal items, like keys and phones have disappeared from that room, and show up later in the nearby Butlers Kitchen. Many believe he is the spirit of an unknown transient who was found dead in the room while the castle was abandoned.

In the Butlers Kitchen, next to the Tea Room, it is said spirits mingle when the room is empty. People will hear voices, but when they arrive, no one is there. Others report seeing tablecloths moving, and items being knocked off the tables.

On the other side of the main stairway, is the Chapel. It is a humble little church where Father Francolon and the nuns would hold daily mass. On the wall of the chapel is a large painting of Father Francolon. On the opposite side of the chapel is a small study, where the priest is said to have done much of his work.

In the chapel, people report cold spots and occasionally an icy hand touching them. On investigations, we have picked up a moving EMF reading which would wander the room and retreat to the study where it disappeared. People have picked up strange anomalies on camera in that room.

Going upstairs from the main lobby is the main hall - a large room where gatherings would take place. On the right is the greenhouse and guest room, which served as the main ward of the Sanitarium. The greenhouse overlooks the front of the castle. It sits immediately above the chapel.

In the greenhouse, you get an overwhelming feeling of hundreds of souls washing over you, and a keen feeling of being watched from the little alcoves and closets which surround the 16-sided room. Many people get feelings of overwhelming doom or sorrow in this room, which go away when you leave. EMF readings and other scientific methods have proven there are no electro-magnetic leaks, nor power lines near enough to the room to cause such feelings. Some have reported being touched with an icy hand, or hearing disembodied voices in the room. Others swear they see something moving out of the corner of their eye. At night, people say they see mysterious faces peering in through the reflections of the windows.

Moving past the greenhouse across the hall, are two doors which lead to Mrs. Francolon's apartment. It is a quaint three room section which includes a bathroom, a small sitting room, and a large bedroom. The room has the original 4 post bed and many of Mrs. Francolon's belongings. There is a photo on display in the castle of the mirror in that room. The mirror's reflection shows a woman said to be the ghost of Mrs. Francolon. This room has a rail which guides visitors through the apartment, and the ghost is said to close the gate if anyone opens it. Some people have reported hearing French voices in the room as well.

Just past the Mother's apartment, is Father Francolon's parlor and bedroom [with] an outdoor balcony. People have reported seeing a Native American man wandering this room however no one seems to know who he is, or why he's there.

Back in the Hall, is a stairway leading up to the servant's quarters in the attic. This area was converted into a gift shop and houses the doll room, and Christmas room. As you go up the stairs to the gift shop, you'll find many great souvenirs and books on ghosts of Colorado, as well as Manitou legends. The gift shop narrows into a hallway which spans the entire spine of the house. There is a door to the back of the castle, which leads to a garden and balcony. In the hallway, there are two rooms, the first and larger room is dubbed the Doll Room. This moderately sized room houses dozens of vintage porcelain dolls which are priceless. The window overlooks the entire castle and has a wonderful view of Manitou Springs. The next room is called the Christmas room, which has many Christmas decorations, trees and other novelties. The spirit of a small girl resides in these rooms. There have been many accounts if this active little spirit, who seems curious about the people who visit her.

She has been spotted on dozens of occasions by guests who think she is a lost visitor, but she vanishes when they get close. The gift shop staff report seeing her often from the end of the hall, which they can clearly see from the cashier area. She also seems to love communicating with paranormal investigators, who give her the most attention. Every time I have investigated Miramont, she is always the most active spirit. Once she discovered a way to communicate via a laptop keyboard, and again by manipulating the focus on a camera. There is a very clear EVP of her asking "What's in the living room?" This seems to corroborate the story she is the child who died there while it was an apartment.

Outside of the attic, on the side of the mountain is a garden on the balcony. It is said that another Native American boy resides there as a beggar. No one knows his story, or if he died there. From the garden, you can follow the stairs down back to the Tea Room.

Miramont Castle seems to be a focal point for lost souls, as many of the spirits who reside there did not die there, which is a rarity for any haunted location. It is possibly one of the most haunted places in the country, and certainly one of the most haunted in Colorado.

If you ever want to visit there, you are welcome anytime throughout the year. There is an admission fee, however once you are inside you are free to wander at your own pace. Due to negative response from some ghost investigations and the overwhelming popularity of the castle as a haunted site, the staff has limited the number of official ghost hunts. They ask you do not disrupt the ambiance of the castle with ghost hunting gear or bother the ghosts, as they are residents of the castle.

During your visit you will probably have a personal experience, due to the sheer number of hauntings the castle has. It is a great place to visit, and the rich history alone is worth the trip. If you wish to discover the ghosts at Miramont, mid-afternoon is the best time to find them.

Cheesman Park in 2011 – Photo by Leon Duran

Cheesman Park

Denver

Located in downtown Denver, Cheesman Park was first an Arapaho Indian burial ground, before becoming the Mount Prospect Cemetery in 1858. It was eventually subdivided into sections for the Masonic / Odd Fellows, Catholic, and Jewish burials. The Mount Prospect section was renamed City Cemetery and eventually became the final resting place for Chinese, Union Civil War veterans, and the destitute.

On March 30, 1859 Jack O'Neill was the first person buried there. He was reported to have been shot during a gun fight. Close to 15,000 people were eventually interred there before the city of Denver petitioned Congress for permission to use the space as a city park, which was granted in 1890. It is still unknown as to why the decision was made to convert the land from cemetery to city park - except for the decline in burials after the opening of Riverside Cemetery in 1876.

By 1893 all burials had stopped with the exception of the Mount Cavalry (Catholic) section, due to an injunction initiated by the Denver Diocese. Mount Cavalry remained in operation until 1908 when Catholic burials started at Mt. Olivet Cemetery in Wheatridge. The city of Denver eventually ordered that all of the remains be relocated from City Cemetery. Close to 800 graves were relocated by family members to either Riverside or Mt. Olivet. The remaining bodies were to be removed by a contract undertaker. This caused a series of scandalous reports from desecration to grave robbing, although there is nothing to confirm the accuracy of these allegations. Approximately 9,000 remains were relocated before the city ordered the removal be stopped.

Today, there are an estimated 4,000 bodies still buried under what is now Cheesman Park. The Denver Botanic Gardens sits where the Mount Cavalry section was

Denver Botanic Gardens
Photo by Leon Duran

located. Several residential neighborhoods were built on top of the outer-lying areas of the former cemetery as well.

Construction projects throughout the years have uncovered human remains to include a casket discovered under the foundation of the Botanic Gardens that had managed to become vertical due to the settling of the land.

Reports of paranormal activity in this area are numerous and frequent. One of the first reports was from a man named Jim who had been hired to relocate the graves. Legend has it he was relocating valuable items from the graves into his pockets instead of relocating the bodies when he felt what he described as a ghost landing on his back. He left the job site that day and never returned.

People report voices and moans inside the park as well as in the houses and businesses surrounding it. Residents have reported spirits in the windows of their homes that vanish on closer inspection and misty, human-like figures can be seen roaming the park.

On moonlit nights, people claim to be able to see the outlines of the graves and some investigators report feeling like they're being held down if they lay in the grass at night. The park closes at 11 PM, but there are still a few nighttime hours in which the extremely brave can venture into the park to check it out!

Chillers Bar and Grill

Loveland

Located at 128 E. 4th Street in Loveland, the Chillers Bar and Grill is home to some strange and unusual activity. Paranormal investigator and founder of the Rocky Mountain Ghost Hunting Club, Jim Beard tells of the resident spirit affectionately named "George" by the staff. He has been blamed for everything from temperature fluctuations to miscellaneous items being thrown across the bar.

> "I started hanging out there several years ago when it was Ragyn Ryans. I started hearing stories about their resident ghost there that they call George. Apparently he was stabbed or killed back when it was the Buck Horn Saloon. He's been known to throw stuff across the bar such as ice scoops which was witnessed by one bartender. Glasses seem to

fall off the bar. One afternoon I was sitting there talking to one of the bartenders and a glass fell off. We looked at each other and both said oh that was just George…I have checked with Loveland Police Department's record division and they have no report of any stabbings or shootings [at that] address. So unless someone comes forward we may never know the true story. But until then I get many people that come up to me at night that went to the back room or to the restrooms in the back and their faces look white and pale. I ask them what's the matter and they say they saw George sitting back there. I ask them what did he look like and what was he wearing and the answer always the same. I tell them well you just saw George hanging out in his usual spot."

Chillers is also the current headquarters for the Rocky Mountain Ghost Hunting Club. Investigations can be scheduled through Jim Beard jimco04@yahoo.com

Outlaws & Lawmen Jail Museum – Photo by Frank Copley

Outlaws & Lawmen Jail Museum

Cripple Creek

Frank Copley, Contributor
Additional contributions by Nancy Byers, MPPIR and Michelle Rozell, Museum Manager

The discovery of gold by cowboy Bob Womack near Cripple Creek brought both problems and riches for Colorado and Cripple Creek. The small ranching community grew from a mere 37 people to a mining district of over 50,000 in just a few years. This presented challenges to a local government seat that was a mere 20 miles as the crow flies away in Colorado Springs, though on the other side of the 14,000ft mountain of Pikes Peak. The long traveling distance around Pikes Peak led to local complaints about taxes going mostly to Colorado Springs, and law enforcement and other officials who were taking too long to respond to the needs of the camp led to the forming of Teller

County. In a late night session on March 23, 1899 the Colorado Congress split off the western mountainous portion of El Paso County and a piece of northern Freemont County to form Teller County.

The influx of all sorts of persons from the agents of investors, miners, shop keepers and others looking to make their fortunes, also brought the scam artist, card sharks, prostitutes and other sordid individuals to the area. Teller County would need a jail to house those that would not follow the law or those who just got out of hand and needed to sleep off a night of rowdiness.

Believed to have been commissioned shortly after Teller County was formed, construction began on the Teller County Jail and

Main cell block along the west wall.
Photo by Frank Copley

was completed in 1901. The jail is a simple red brick block building of the 1890's western style. The jail houses a cell block of fourteen cells, ten on the lower level and four on the upper level in the main part of the building. In the front of the building upstairs from the offices is a small section of three cells that was used for the women's jail and juvenile detention during early operation.

In addition to housing criminals of all kinds, the jail also housed the insane, served as an orphanage, a morgue for unclaimed deceased and a clinic during the Spanish flu of 1918. Times changed in the late 1980's and the jail's lack of an appropriate recreational area for the inmates led to its closure. The Teller County jail in Cripple Creek was officially closed in 1992, as a result of a lawsuit brought by the ACLU against Teller County. The building remained opened only as a temporary holding area for detaining criminal suspects until 1996. The new Teller County Jail was completed in Divide, Colorado at that time and the building was left abandoned.

With Cripple Creek in decline and only one major mine in operation, the once booming district suffered. The jail stood empty, the roof partially collapsed, and the electricity disconnected. With limited stakes gambling bringing much needed revenue from the gamblers, the jail was reopened as the Outlaws and Lawmen Jail Museum in 2007.

Mountain Peak Paranormal Investigations and Research (MPPIR), with the help of many of the local historians, have spent many days and nights researching the jail, both its history and its ghosts. The research by MPPIR and local historians have led to the discovery of the lost history of some of the facility's inmate deaths and thus uncovering clues to the ghosts that still seem to wander the jail.

The first known death in the facility is yet unconfirmed, but the traces of this death can be seen on the floor along the east wall of the cell block. A large blood stain that is still seen with the help of a UV light, lays half way down the east walkway. It is believed that a man of Italian decent was either pushed or jumped to his death from the second level of the cell block. This is a local story, and as of now, no written record as of yet has been found concerning this death.

MPPIR's research into the bloodstain and the local story has had some results. From the paranormal research, using digital recorders the sound of heavy breathing has been recorded near the blood stain. With incomplete records, at this time we have not discovered his name or occupation of the inmate, only that he may have been an Italian man.

The second known tragedy in the jail was uncovered by Michelle Rozell, the museums manager and historian. This is the horrible tragedy is of a young woman in her 30's by the name of Olga Knutson.

Olga was a recent mother of an infant son of only 8 months of age, when she fell into a state of dementia. The poor woman was completely distraught and rambling on about unfounded fears of her neighbors and an alleged threat from a Sheriff to hang her. Olga also feared for the health of her newborn son. The local doctors could find nothing wrong with the infant and tried their best to confer this to the new mother.

Olga's condition only worsened over the next few days. She refused food, drinks and did not sleep, most likely causing her condition to worsen. She was remanded to the Red Cross Hospital, wherein she became violent towards the staff. The violent outbursts only led to her being charged with Insanity by the court. Olga was deemed insane and remanded to the Teller County Jail.

In the jail Olga was given a sleeping potion and only at that time did she finally succumb to sleep. The combination of lack of sleep, nourishment, insanity and possibly the sleeping potion seemed too much for the now frail Olga. At 5 o'clock (am?) it appeared to the matron that Olga's condition had worsened. The local physician was summoned too late, as Olga Knutson expired before he arrived.

Mountain Peak Paranormal Investigations began research at the Jail Museum in 2009. Their findings have been the major reason for the site being [considered] haunted, outside of the locals already having this opinion.

Stories from residents of Cripple Creek about the jail facility after its closing, until 2007 when it reopened, have ranged from seeing lights in the building and seeing people moving around inside. Local police would be called to these reports and find nothing but locked doors and no signs of entry. These events all occurred during the time when the building was abandoned, locked and without electricity.

The opening of the building as the Outlaws and Lawmen Jail Museum did not stop ghosts from wandering its halls. On more than one occasion, the Museum's alarm has sounded with the police arriving to find the building secured and undisturbed.

On one occasion it was reported that a little girl wandered in the front door right up to a visitor to only slowly disappear. Two different times, an apparition of a woman in clothing from the early 1900's has appeared and then simply vanished.

A museum employee told of an older man, dressed in what appeared to be a police uniform standing at the front door waiting to come in. Since it was after closing and the door was locked, the employee went to the door to greet the man she thought would be a local officer. [The officer] was gone. The museum employee looked out the front door only to find the street vacant. She did state that it was odd that he would be dressed in a uniform that appeared to be from a much older time.

Mountain Peak Paranormal has researched and continues to do paranormal research at the old jail. Our research team has seen shadows of men moving about the main cellblock. Even the footsteps of what is believed to be the night watchmen have been found on audio recordings by the MPPIR team. On one occasion that this writer was present for, a laser trip-sensor was set off on the second level with several witnesses present and with no one present on the second level of the cell block.

Disembodied voices have been heard and recorded by the MPPIR research team on many investigations. Some of the things they have said include "Stick it out!" and "Get Out!" being heard in the solitary confinement cell.

Disembodied voices of children have been heard laughing in the main cell block and a woman having a one-sided conversation with someone have also been heard in the main cell block.

Electronic voice phenomenon (EVP) on recordings in the main cell block can be eerie. One such is that of the name "John" sounding like it is being screamed right near a video camera recording in cell #5. There was a John present nearby but neither he nor the other investigators present heard the name yelled out.

A female voice has been recorded but not heard by the investigators on the 2nd floor in the woman's jail area above the offices. The recordings include a woman stating "Its Amie",

Women's Cell #2
Photo by Frank Copley

and another time, with the same voice pattern, it seemingly responds to the investigator saying, "It's cold in here".

In May of 2011 a recording in the woman's section of the jail there was a female voice sounding very much like the same as the other two mentioned above. The recording was a full sentence saying, "Ok, you can come out now." This was recorded on a digital video camera in the Matron's room. The voice was recorded after an investigating group left one of the cells on the 2nd floor and entered the Matron's room at the end of the hall. The voice sounds as if it is in the hallway area outside the room.

We do not know who this Lady is. Was she a matron? Was she an inmate? Whoever she is, she seems very protective. Some have said she is protecting the children that once stayed within the jail's walls.

The last place [to look at] on this short trip is the basement. It is regrettably closed to the public but MPPIR has been allowed access for paranormal research.

The basement area towards the rear of the building is a wide open, almost warehouse-looking area with 2 doors at the rear of the building. One door on the east side of the south wall is heavily barred, and its original purpose may have been to bring prisoners in from the rear of the building. That leaves us wondering about the second door. This door is also reinforced but is large enough to bring a carriage or small car through. This area is known for its shadows moving in the darkness.

By far though the most interesting and eerie part of the basement is located down stairs from the entryway above and located directly below the offices. This area once housed a coal fired steam boiler, a coal room, and the "Ice Room".

With no door, only an archway, entry into the "Ice Room" is easily accessed by those wishing to see, by far, the eeriest room of the basement area. Used to store ice cut in the winter, it also doubled as a morgue for unclaimed deceased. Entry by some into this room has led to their hasty retreat back out of this unassuming area. Nothing in the bare rock walls and bricked up window would lead you to believe it was anything but a simple storage area. It is just the feeling of dread that many receive while in the room.

The "Ice Room" has given electronic voice phenomenon that bothered investigators the most. One such EVP, "Come join me in death!" was told to MPPIR's founder and lead investigator, Ralph.

Batteries will drain, cameras will fail to take pictures and video cameras have been known to shut off on their own in this room. Every investigator that has entered has been heard complaining about equipment failure in the "Ice Room."

Do the departed that had to stay in the "Ice Room" during the winter, awaiting the ground to soften so they may be buried, not want to be photographed or recorded? Do they just want us to leave them in peace in this room? Or as one voice stated, do they want us to join them?

The Jail Museum is a piece of American history. Being well preserved with its cells left in their original condition, complete with the graffiti from those that came and went from its halls; this is a place to visit. Bring along some recording equipment. You might be surprised what you hear when you play it back later.

Frank Copley is the Blog Administrator and Photographer for the Mountain Peak Paranormal Investigations and Research (MPPIR). Please visit their website: www.mppir.com for more information.

Stanley Hotel – Photo by CCPI

Stanley Hotel
Estes Park

In 1903 Freelan Oscar (F.O.) Stanley and his wife, Flora came to Colorado seeking relief from the Tuberculosis that was deteriorating the health of the Stanley Steamer tycoon. With his health improving and having fallen in love with the Estes Park area, F.O. Stanley chose to invest his money in the town by developing a luxury hotel where people from the east could come to relax and take in the majestic Rocky Mountains.

Stanley purchased 160 acres of land from Windham Thomas Wyndham-Quin, 4th Earl of Dunraven, simply known as "Lord Dunraven" after the mass settlement of the area placed a damper on plans for a game preserve and of Dunraven becoming the Earl of Estes. There are also conflicting reports about the legality of Lord Dunraven's land ownership compared to his immigration status that perhaps forced him to sell as well.

Construction of the luxurious hotel began in 1907 with the main building being completed in 1909. No expense was spared with

**The lobby of the Stanley Hotel –
complete with a Stanley Steamer**
Photo by CCPI

his accommodations that featured running water, telephones, and electricity. Initially intended as a summer resort, the Stanley Hotel had no heat and catered only to the wealthy. A second building - smaller in scale, but identical in architecture and interior design, known as the Manor House was built for bachelors to keep them separate from single women and families staying in the main hotel facility. Both buildings featured billiard and dining rooms, and only the manor house is said to have had heat so that staff could stay during the winter as caretakers.

A concert hall and carriage house (for Stanley Steamers) was later added. The Stanley's sold the hotel in 1926 and the facility operated only during the summer months until the 1980's after water heat had been added to the main building, although the hotel is reported to have closed briefly during WWII.

During a 1974 visit to the hotel, novelist Stephen King was inspired to write his classic *The Shining*. Some people believe he wrote the novel while staying in the hotel, but hotel staff report that he was actually living in Boulder at the time and only received inspiration for the story while staying in Estes Park.

The 1980 Stanley Kubrick thriller *The Shining* utilized a Hollywood set for the famous Overlook Hotel, but the 1995 miniseries was filmed at the Stanley with Stephen King in attendance. The 1994 comedy *Dumb and Dumber* also utilized the Stanley Hotel as the set for the Hotel Danbury.

In addition to the Hollywood stamp of approval and rich history, the Stanley Hotel has one feature that just cannot go unnoticed. It's haunted. Reports of ghostly activity range from apparition sightings throughout the hotel to disappearing jewelry with doors opening and closing in room 401, ghostly children running and playing in the fourth floor corridors, piano music being played (presumably by Mrs. Stanley) in the music room, and other numerous reports throughout the property. F.O. Stanley and Lord Dunraven have been identified on several occasions by staff and guests, often having vanished before their very eyes.

On Halloween night in 2006, a television show that was gaining in popularity visited the Stanley Hotel for a live 6-hour televised investigation. That episode of *Ghost Hunters* (SyFy Channel) created a frenzy of interest by the millions who were tuned in that night. Viewers sat glued to their television screens as they listened to a disembodied voice in an underground tunnel and watched as a table appeared to jump on its own in the Manor

House. Other paranormal television shows soon followed suit to include the UK show *Most Haunted* (Travel Channel) in 2008 and *Ghost Adventures* (Travel Channel) in 2010. It seemed everyone wanted to capture a glimpse of the Stanley Hotel's permanent residents.

A 2008 photo of the author exploring an area of the Stanley Hotel known for disembodied voices

Photo by CCPI

Amateur and experienced paranormal investigators now flock from near and far to explore the famous Stanley Hotel and paranormal conferences are regularly held where the public can investigate with the television celebrities.

The Stanley Hotel also offers guided tours where visitors can walk through the facility while learning about the rich history of F.O. Stanley, Lord Dunraven and the ghostly inhabitants. These tours take place several times every day, and can fit nicely into tourist schedules. Private investigations are also offered with the assistance of a highly skilled, knowledgeable, and dedicated hotel investigator, but advanced reservations are needed.

For more information or to make reservations, contact the Stanley Hotel 800-976-1300 or www.stanleyhotel.com

Potter's Field
Grand Junction

The Potter's Field cemetery is reported to be the very first cemetery in the Grand Junction area. It is located south of the intersection of B 3/4 Road and 26 3/8 Road next to the Grand Junction Police firing range.

The earliest known burial there was Baby Garland in 1881 with Julian Pacheco being the last known to be buried in 1936. There are only two actual headstones and several wooden crosses to mark the rapidly deteriorating graves. While there are only eight known graves inside the fence, it is speculated that there could be as many as eighty.

The reason for this speculation is due to legends about trigger-happy law enforcement in the early twentieth century who believed that the "criminals" should not be buried in consecrated ground, i.e. the Orchard Mesa Municipal Cemetery located on the hill above. The number of unmarked graves continues to climb as more are discovered.

During a training session on April 14, 2009 the Colorado Coalition of Paranormal Investigators (CCPI) discovered a casket

handle near a sunken grave in the cemetery. Until that point, there had not been any evidence of a grave at that location. The handle has since been replaced, the sinkhole filled, and a ring of river rocks marks the newly-discovered grave.

A casket handle found near a sunken grave.
Photo by CCPI

Grave discovery and identification aren't the only things being researched at the old cemetery. Potter's Field also has a reputation for being very haunted! From cryptic and sometimes violent EVPs to full-bodied apparition sightings, Potter's Field is a place where investigators go to hone their skills or recreate previous experiences. Some investigators even believe they are able to identify which spirits are active and when.

Some of the residents of Potter's Field that investigators claim to have contact with on a regular basis include Elmer Price, Mary

Middaugh, and Bertha Kaufman. Bertha's grave is a concrete vault where visitors leave toys, flowers, and money for the little girl. Investigators report having pant legs tugged on by a small child, and seeing what appears to be a youngster playing peek-a-boo with them in the bushes near her grave.

Mary Middaugh is buried with Mrs. A.A. Middaugh (presumably her mother) in one of the unmarked graves. Her circumstances are unknown and the date of death is listed as 1913. Investigators have gathered EVP of a young female voice near the back part of the cemetery where Mary is believed to be buried.

Elmer Price is reported to be buried in one of the unmarked graves as well. His birth date is unknown and his date of death is only listed as 1911. Investigators believe that interactions with Elmer result in violent or offensive EVP and he is frequently blamed for physical attacks that are very rare at the cemetery.

Regardless of the spirit that decides to interact, Potter's field is a frequent paranormal pleaser that should be checked out at least once. Standard cemetery rules apply though, so all investigating needs to be completed by 10 P.M. and the area is patrolled by the Grand Junction Police Department.

Stone Monument to Doc Holliday – Photo by CCPI

Doc Holliday's Grave / Kid Curry's Grave

Linwood Cemetery

Glenwood Springs

John Henry Holliday was born on August 14, 1851 in Griffin Georgia. When he was twenty years old, he graduated from the Pennsylvania College of Dental Surgery, earning him the infamous nickname "Doc".

Doc traveled all over the country, satisfying his gambling addiction and developing a reputation for being a deadly gunman. It has been said that he felt gambling was more profitable than practicing dentistry, as most of his patients were wary of his cough from Tuberculosis.

He met Wyatt Earp some time during 1877 and the two developed a very strong friendship. They, along with Virgil and Morgan Earp, made a stand against a cattle rustling gang called the Cowboys at the OK Corral in Tombstone, AZ on October 26, 1881. The fight lasted 30 seconds and resulted in the deaths of three Cowboys. That conflict went down as the most famous gunfight in the history of the Old West.

After the July 14, 1882 shooting death of gunslinger Johnny Ringo, Doc moved to Colorado where he reportedly spent the rest of his life. He came to Glenwood Springs where the altitude was easier on his lungs. He wanted to partake in the medicinal hot springs, but ultimately spent the last months of his life in the Hotel Glenwood Sanatorium, suffering from the alcohol and laudanum addictions that resulted from treating his Tuberculosis.

Doc Holliday died on November 8, 1887. He was 36 years old. The nurses reported his last words being "Now that's funny" although historians debate that due to his addiction to laudanum, saying that coherent communication was unlikely. He died a pauper and was buried in the Potter's Field section of the Linwood Cemetery in Glenwood Springs. A wooden cross was erected to mark his grave, but has since deteriorated. Cemetery records were lost or destroyed sometime between 1887 and 1939, leaving no way of knowing the exact location of his body. In the 1950's a stone monument was erected to indicate his presence within the cemetery. In 2004, the stone was replaced with a restored headstone from the late 1800's.

Doc Holliday isn't the only significant figure from the Wild West buried in Linwood. Gunslinger Kid Curry is also somewhere in the potter's field section - his wooden marker having also eroded away. If ever there were a spirit to have reason to return and haunt, it's Curry.

In 1867 Harvey Logan was born in Richland Township, Iowa. After the death of his mother, he and his brothers went to live with their aunt in Missouri. They earned a living working with horses, but were known to be reckless with whiskey and women. He and his brothers adopted the name Curry for themselves after friend, "Flat Nose" George Curry.

Kid became an outlaw after a misunderstanding with a neighbor (and subsequent murder) led to him to flee, fearing an unfair trial. He rode with notorious outlaws like Black Jack Ketchum and later Butch Cassidy. He was said to be one of the deadliest gunmen of the era.

In 1904, Kid went to work for the railroad in Glenwood Springs under the alias J.H. Ross. It is thought that he took the job to learn when the payroll trains came through. On June 7, 1904 Ross robbed a train just outside of the town of Parachute. He was wounded by the posse that pursued him and shot himself in the head to avoid going to jail. A death photo was taken prior to his burial in the Linwood cemetery and sent to various law enforcement agencies who speculated that J.H. Ross was in fact, Kid Curry.

To verify the identification, a physician was sent to exhume Ross' body. Various physical features and scars were noted, confirming the suspicions that J.H. Ross was actually Harvey Logan, a.k.a. Kid Curry. Notification of Curry's death was sent to the Pinkerton's who had been on his trail for years. Pinkerton agent, Lowell Spence was sent and once again the body was exhumed for identity confirmation. Could having your body buried and exhumed on three separate occasions result in a less-than peaceful eternal rest?

The Linwood Cemetery sits atop the Pioneer Cemetery Trail in Glenwood Springs and overlooks the town. There are over 500 people buried on the hill, although there are only records for 358. In 1939, heavy rains inundated part of the cemetery, resulting in stories about bodies and caskets being washed down the hill. The stories tell of the bodies being re-buried in the cemetery with no

A worn headstone leans next to the eroded trail

Photo by CCPI

real way of knowing if the correct people were placed with their respective headstones. Logistically, this is impossible. The worst of the damage would have been a few relocated headstones. Due to the depths at which bodies are usually interred, no actual bodies could have washed down the hill.

Reports of paranormal activity range from shadows to full-bodied apparitions. EVP gathered in the cemetery are usually male voices, although the occasional female voice has been reported.

The Frontier Historical Society holds an annual "Ghost Walk" event at the Linwood Cemetery. For three weeks every October, guests can purchase tickets and make the ½-mile hike to the top by lantern light to interact with costumed actors portraying Doc Holliday, Kid Curry, and others buried on the hill. The event is historical and not intended to be paranormal, but is well worth the money and exercise to experience a piece of history.

To learn more about Doc Holliday, Kid Curry, and the town of Glenwood Springs, please visit the Frontier Historical Museum: 1001 Colorado Avenue. For more information on the Linwood cemetery or the Ghost Walk, contact the Frontier Historical Society (970) 945-4448. www.glenwoodhistory.com

Orchard Mesa Cemetery Main Entrance – Photo by CCPI

Orchard Mesa Municipal Cemetery

Grand Junction

Sitting on a hill overlooking the confluences of the Gunnison and Colorado rivers is the Orchard Mesa Cemetery. It is the largest cemetery in the Grand Junction area with nine sections dividing the massive burial ground that has been in operation since the early 1900's.

On a hill directly south of the cemetery is Crawford's Tomb. For decades, Crawford's Tomb has been an attraction for thrill seekers and investigators alike. Vandals however have desecrated the final resting place of Grand Junction's founder, George A. Crawford and a fence now prohibits access to the monument.

Paranormal activity inside the cemetery ranges from apparition sightings and amazing EVP recordings to people being touched or chased and even shape-shifter or skinwalker type encounters.

One such encounter was reported by the Colorado Coalition of Paranormal Investigators (CCPI) during a 2005 investigation. The team had been preparing to practice inside the cemetery when a wild fox approached them outside the cemetery walls. The team decided to take a few photos of the animal, but it would not appear in any of their digital photos. Two video cameras also malfunctioned while trying to capture images of the fox – recording only audio and no video. Finally, they were able to capture the animal with a Polaroid instant camera, however when the picture developed, all that could be seen were its eyes.

Since the CCPI encounter, others have reported seeing and having similar encounters with animals in the cemetery to include foxes, coyotes, and mountain lions. The most active section is reported to be the Veteran's section. The Veteran's section was the first section of the cemetery to be developed in 1898. The other eight sections were built around it, before the City of Grand Junction took over in 1978.

The cemetery is open year-round until 10 PM and is patrolled by the Grand Junction Police Department. Rain or shine, Orchard Mesa is sure to please – if you know where to look!

Riverside Cemetery – Photo by Brandy Nelson

Riverside Cemetery

Denver

The Riverside Cemetery is Denver's oldest operating burial ground. It was founded in 1876 – the same year Colorado gained its statehood and the United States celebrated 100 years as an independent nation. Located at 5201 Brighton Boulevard, it is 77 acres of park-like resting places for over 67,000 people. It is said that only half of the known people buried there have markers. This is in part due to the relocation of remains from the former City Cemetery, now known as Cheesman Park.

Some of the most notorious and influential Colorado residents are buried in Riverside to include Governors, Mayors, and Civil War veterans. Negro League third baseman and shortstop, Oliver "Ghost" Marcelle is listed as being buried in an unmarked grave in Riverside, as well as Clara Brown, the first African-American woman to cross the Great Plains during the gold rush. Colorado mining pioneer, Lester Drake rests in the cemetery with a stone replica of his cabin marking the spot.

A replica of Lester Drake's cabin
Photo by Brandy Nelson

Investigator Brandy Nelson has reported activity at Riverside Cemetery ranging from rocks being thrown at people to having hats taken off of heads and items forcibly removed from investigator's hands.

In 2001, Riverside lost its water rights from the South Platte River leaving the grass and trees victims of the droughts of decades past. Plots are no longer being sold, but people owning previously purchased spaces can still be interred there.

If you are interested in helping this priceless piece of Colorado history, you can contact the Friends of Historic Riverside Cemetery via email: volunteers@friendsofriversidecemetery.org

Contributors

Balzano, Christopher: 7-11

Beard, Jim: 63-64

Byers, Nancy: 65

Copley, Frank: 65 - 74

Duran, Leon: 60, 61

Martin, Zachariah: 21-22

Martinez, Jack: 51-59

Nelson, Brandy: 89, 90

Rozell, Michelle: 65, 68

Colorado Coalition of Paranormal Investigators (CCPI)

Mountain Peak Paranormal Investigations & Research (MPPIR)

Clarissa Vazquez is a Colorado native and avid paranormal investigator, having founded the Colorado Coalition of Paranormal Investigators (CCPI) in 2004. She has written several books on ghosts and the paranormal to include *Ghosts of the Heart: A Paranormal Investigator's Journey, Ghosts and the Bible, and No Monsters Here!* In addition to her own literary works, Clarissa has been a featured author in books such as *Picture Yourself Capturing Ghosts on Film*, by Christopher Balzano, *Ghosts from Coast to Coast*, by Kalyomi, and Ghosts *of Colorado*, by Dennis Baker.

She is most noted for her paranormal research and development of the Phantom Hitchhiker Project – designed to prove that paranormal investigation *can* be conducted scientifically and thus should be recognized as legitimate research by the mainstream scientific community.

Clarissa has been featured on terrestrial and Internet radio shows alike to include *Coast to Coast A.M.* with George Noory, *Spooky Southcoast* with Tim Weisberg, and *Darkness on the Edge of Town* with David Schrader. You can sometimes catch her hosting or co-hosting radio programs *Para-Scope Uncensored* and *Crossing the Void Radio.*